I M A L S

WITHDRAWN

TIGERS

BY VALERIE BODDEN

CREATIVE EDUCATION • CREATI

Published by Creative Education and
Creative Paperbacks
P.O. Box 227, Mankato, Minnesota 56002
Creative Education and Creative Paperbacks are
imprints of The Creative Company
www.thecreativecompany.us

Design and production by The Design Lab
Production by Rachel Klimpel
Edited by Alissa Thielges
Art direction by Rita Marshall

Photographs by 123RF (Isselee), Alamy (Peter Maszlen,
surasak suwanmake, Suretha Rous), Getty (Schafer &
Hill, toni), iStock (GlobalP, Puttachat Kumkrong, Sourabh
Bharti, Vinod Bartakk), Shutterstock (Dennis van de
Water, Eric Isselee, Martin Mecnarowski, Nachaliti,
slowmotiongli, Vaclav Sebek)

Library of Congress Cataloging-in-Publication Data
Names: Bodden, Valerie, author.
Title: Tigers / by Valerie Bodden.
Description: Mankato, Minnesota: The Creative
Company, [2023] | Series: Amazing animals | Includes
bibliographical references and index. | Audience: Ages
6–9 | Audience: Grades 2–3 | Summary: "Meet the
largest wild cat in the world! These wild cats rule the
Asian jungle and are powerful predators. This book
explores their features and behaviors, including hunting
prey and raising cubs. A folktale explains how the tiger
got its stripes"—Provided by publisher.
Identifiers: LCCN 2021045220 (print) | LCCN
2021045221 (ebook) | ISBN 9781640265530
(hardcover) | ISBN 9781682771082 (paperback) |
ISBN 9781640007512 (ebook)
Subjects: LCSH: Tiger—Juvenile literature.
Classification: LCC QL737.C23 B6434 2023 (print) |
LCC QL737.C23 (ebook) | DDC 599.756—dc23
LC record available at https://lccn.loc.gov/2021045220
LC ebook record available at https://lccn.loc.
gov/2021045221

Table of Contents

Tigers are the only big cats that have striped fur.

Tigers are big cats. They live in the wild. There used to be nine kinds of tigers. But some of them died out. Today, there are only six kinds of tigers left. They are all endangered.

endangered a plant or animal that has very few living members and could die out completely

Tigers' big paws help them swim well.

Tigers are normally orange with black stripes. Sometimes tigers are white with black stripes, but this is rare. All tigers have white fur on their bellies, throats, and legs. Tigers have big teeth. They have long, sharp claws.

rare not common or usual

Tigers are the biggest cats in the world. If a tiger stood on its back feet, it would be taller than a grown-up human! Male tigers can weigh more than 500 pounds (227 kg). Female tigers are smaller.

Each tiger has a unique pattern of stripes.

Tigers that live in snowy places have thick fur.

Tigers live in forests in Asia. Some forests are cold with a lot of snow. Tigers that live there have thick fur to stay warm. Other tigers live in hot swamps. They cool off in the water. Tigers are good at swimming.

swamps areas of land that are wet and have lots of trees

Tigers sneak up and pounce on their prey.

All tigers eat meat. They are powerful hunters. They eat big **prey** like deer and pigs. Some tigers hunt monkeys, too. Tigers do not usually hunt people. But they will attack if a person gets too close.

prey an animal that is killed by another animal for food

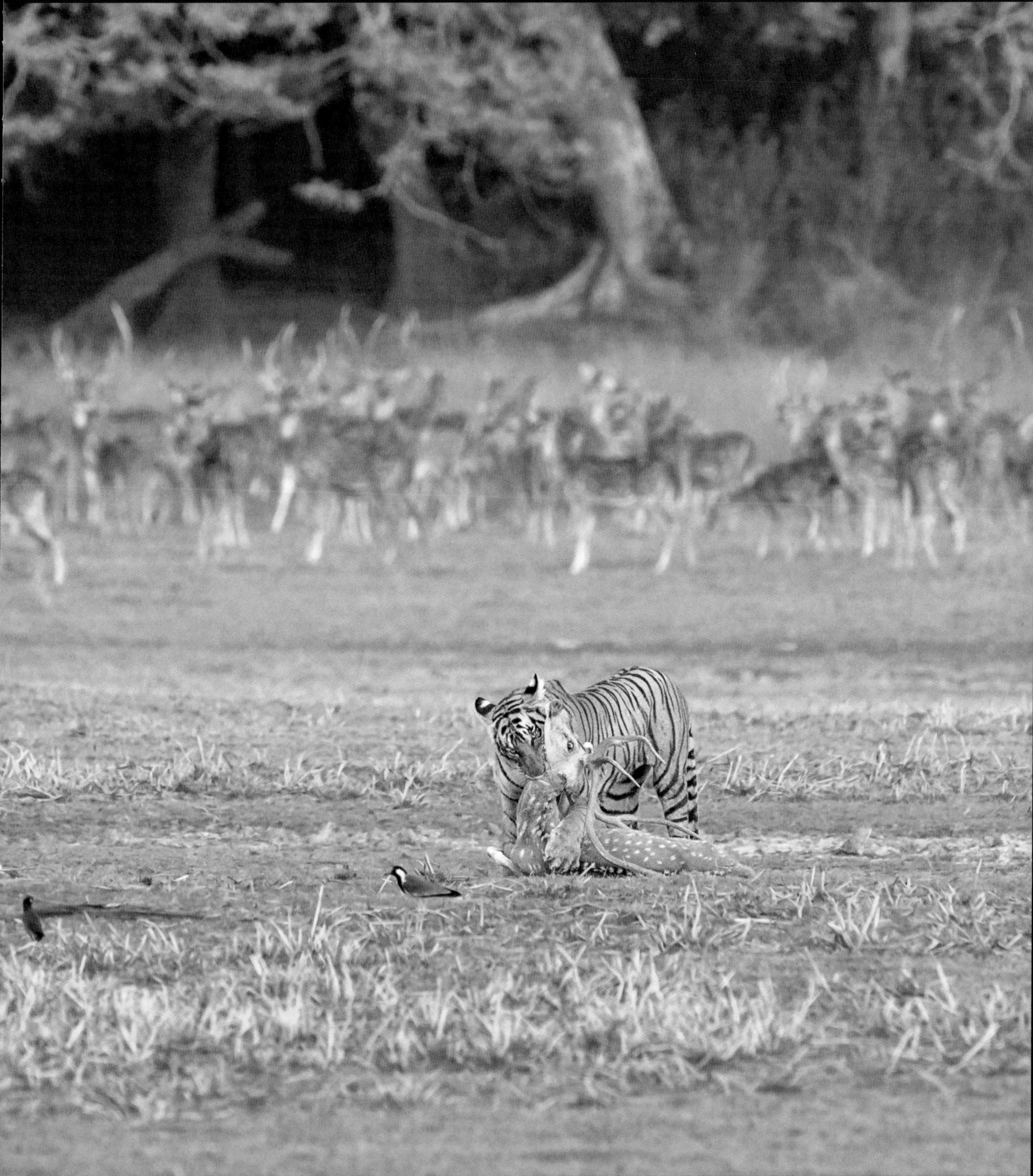

Tiger cubs hide while their mother hunts for food.

Female tigers have two to four cubs at a time. At first, the cubs drink their mother's milk. Then she teaches them to hunt. The cubs leave their mother by the time they are three years old. Wild tigers can live for 10 to 15 years.

cubs baby tigers

Grown tigers live alone. They sleep most of the day. At night, they get up. If they are hungry, they hunt for prey. Tigers can see well in the dark.

Tigers sometimes rest in tall grass.

Tigers make different sounds. They can make a quiet "chuff" sound through their noses. They can growl and snarl, too. And they can roar very loud.

Tigers may roar to tell other tigers to stay away.

Today, there are many tigers in zoos. People love watching these big cats eat, sleep, and play. They are amazing animals to see up close.

Many people go to zoos to see tigers and other big cats.

A Tiger Tale

Why do tigers have stripes? People in Asia used to tell a story about this. They said that the tiger asked a man for wisdom. The man had to travel a long way to get the wisdom. He did not want the tiger to eat his goats while he was gone. So he used ropes to tie the tiger to a tree. The tiger pulled on the ropes to escape. The ropes dug into his fur and left stripes on it!

Read More

Grack, Rachel. *Tigers*. Minneapolis: Bellwether Media, 2022.

Murray, Julie. *Tigers*. Minneapolis: ABDO Publishing, 2020.

Williams, Lily. *If Tigers Disappeared*. New York: Roaring Brook Press, 2022.

Web Sites

San Diego Zoo Kids: Tiger Life-Drawing Lesson
https://kids.sandiegozoowildlifealliance.org/activities/tiger-life-drawing-lesson
Learn how to draw a tiger from a movie artist.

Smithsonian's National Zoo: Tiger Questions and Facts
https://nationalzoo.si.edu/animals/news/how-big-are-tigers-and-more-tiger-facts
Zookeepers answer questions about tigers.

Note: Every effort has been made to ensure that the websites listed above are suitable for children, that they have educational value, and that they contain no inappropriate material. However, because of the nature of the Internet, it is impossible to guarantee that these sites will remain active indefinitely or that their contents will not be altered.

Index